Mohamed Ali Jemni
Hamdi Hentati

Alternative Fuels and Hydrogen Issues in Vehicles

Mohamed Ali Jemni
Hamdi Hentati

Alternative Fuels and Hydrogen Issues in Vehicles

Analysis of technical, economic and environmental progress

ScienciaScripts

Imprint
Any brand names and product names mentioned in this book are subject to trademark, brand or patent protection and are trademarks or registered trademarks of their respective holders. The use of brand names, product names, common names, trade names, product descriptions etc. even without a particular marking in this work is in no way to be construed to mean that such names may be regarded as unrestricted in respect of trademark and brand protection legislation and could thus be used by anyone.

Cover image: www.ingimage.com

This book is a translation from the original published under ISBN 978-620-2-54967-7.

Publisher:
Sciencia Scripts
is a trademark of
Dodo Books Indian Ocean Ltd., member of the OmniScriptum S.R.L Publishing group
str. A.Russo 15, of. 61, Chisinau-2068, Republic of Moldova Europe
Printed at: see last page
ISBN: 978-620-4-11003-5

Future and challenges of alternative fuels and hydrogen in vehicles

Analysis of technical, economic and environmental progress

Mohamed Ali Jemni

Hamdi Hentati

Foreword

This book is intended for students in I.S.E.T. technology classes and for undergraduate and graduate students. It is a pedagogical support for the study of the challenges of using alternative fuels and hydrogen in vehicle combustion engines. The aim of the book is to focus on the current problems of fossil fuel use and the future of alternative fuels.

First, the details of conventional and alternative fuels are illustrated. Then the technology of internal combustion engines is presented in a simplified and clear manner. Finally, the techniques for using alternative gases in engines are identified.

Table of Contents

Chapter 1: Fossil Fuels in Vehicles; Economic and Environmental Issues

1. Introduction

The recent surge in oil prices and concerns about global warming have brought the issue of transportation back to the forefront of our concerns. Vulnerability to oil, resulting from its rapid growth over the past 30 years, and its effects on toxic emissions are of concern to most countries.

Therefore, the use of alternative fuels in the land transport sector, such as LPG, CNG and hydrogen, in combustion engines powered by conventional fossil fuels is an appropriate solution to mitigate these nuisances.

2. What is oil, the origin of fossil fuels?

Oil is a mixture of several hydrocarbons and molecules with percentages of sulfur, nitrogen and oxygen.

Oil comes from the thermal decay over millions of years of organic substances, of plant and animal origin, existing in the mother rocks of this constituent.

Under the action of the high temperatures which prevail there, liquid oil is formed accompanied by gas.

Oil and gas flow upward to porous reservoir rocks where they are confined (Figure 1.1).

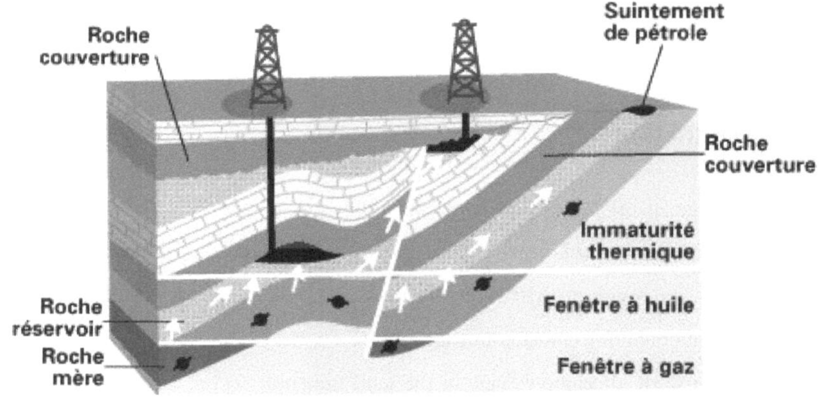

Source: IFP énergies nouvelles

Figure 1.1. Petroleum system

Oil is the most demanded energy source in the world. It covers more than a third of the world's energy needs.

It is the main raw material for fuels used in transportation (cars, trucks, planes).

It constitutes the major part of the raw material of fuels intended for transport (vehicles, heavy goods vehicles, maritime transport, etc.).

Nowadays, the disruption of oil production and discovery estimates (figure 1.2), forces the transport industry to look for other solutions to power internal combustion engines such as alternative gaseous and liquid fuels and hydrogen.

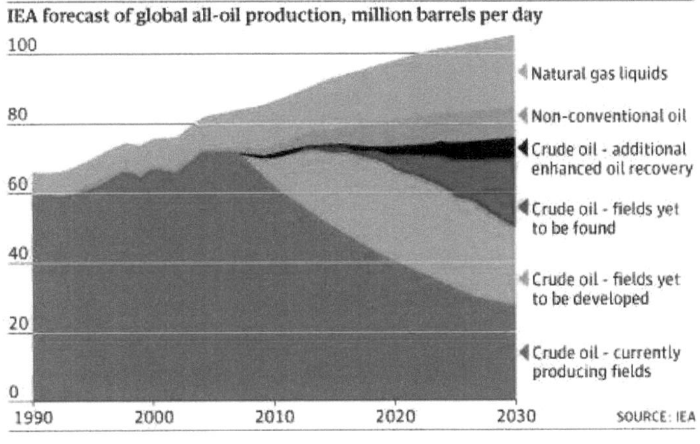

Source: IFP énergies nouvelles

Figure 1.2: Evolution of oil production

3. Conventional Fossil Fuels in Transportation

Fossil fuels in the transport sector are the products of the oil refining process. Their combustion results in a significant amount of heat. Their formulation evolves with the requirements of the engines and especially with the environmental regulations. The use of fossil fuels in transport is responsible for the current toxic emissions of CO, CO_2, etc.

The conventional fossil fuels used in internal combustion engines are diesel and gasoline.

3.1 Gasoline

Gasoline is a product of crude oil refining. It is a mixture of petroleum hydrocarbons and other fuel products or additives that improve the characteristics of the fuel, such as octane rating. Gasoline is liquid and has a high energy density as measured by its calorific value. Oil also contains varying amounts of sulphur depending on its origin. Sulphur must be removed to meet gasoline quality standards. During refining, the sulphur content of gasoline is reduced to meet regulatory standards. The composition of gasoline has changed over the years. Initially to increase performance, but in recent years also to reduce the environmental impact. The environmental impact of using petrol is due to the combustion in the engine but also to its intrinsic characteristics. The general characteristics of gasoline are summarized in Table 1.1.

Table 1.1: Characteristics of gasoline

Features	Regular gasoline	Super Gasoline
Lower heating value (MJ.kg $^{-1}$)	43.3	42.9
Density (kg/l)	0.735	0.750
Calorific value by volume (MJ/l)	31.8	32.4
Ignition temperature (°C)	450	450

3.2 Gas oil, diesel

Diesel is also produced by refining crude oil. It consists of a mixture of hydrocarbons from the distillation of petroleum and other fuel products or additives to improve the characteristics of the fuel. Diesel contains a higher fraction of hydrocarbons than gasoline. It is therefore less volatile and less fluid than gasoline but has a higher energy density. The composition of diesel has evolved over time and will change again in the future to meet new European requirements.

Table 1.2: Diesel fuel characteristics

Features	Value
Lower Calorific Value (MJ/kg)	42,6
Density (kg/l)	0,840
Calorific value by volume (MJ/l)	35,8
Ignition temperature (°C)	260

4. Environmental issues of fossil fuels in the transport sector

The use of fossil fuels in industry is generally responsible for 82% of current CO_2 emissions (coal 35%, oil 31%, gas 16%). It is also implicated in many serious accidents and in water and air pollution of public health concern.

And despite global efforts to reduce the rate of emissions, the amount is still dangerous (Figure 1.3). We also note that greenhouse gas emissions are directly linked to fossil energy resources (Figure 1.4).

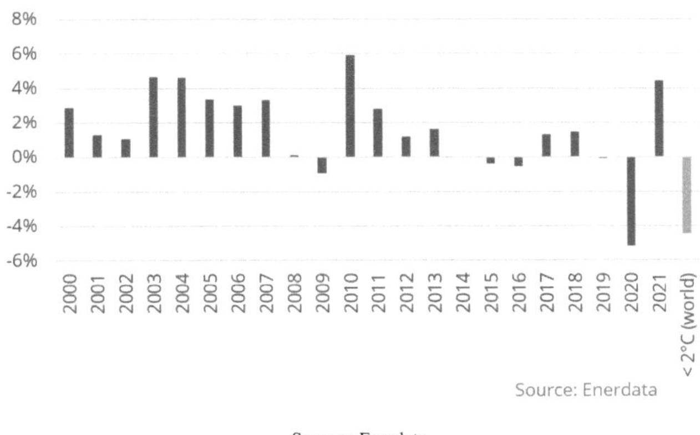

Source : Enerdata

Figure 1.3. Energy-related CO_2 emissions

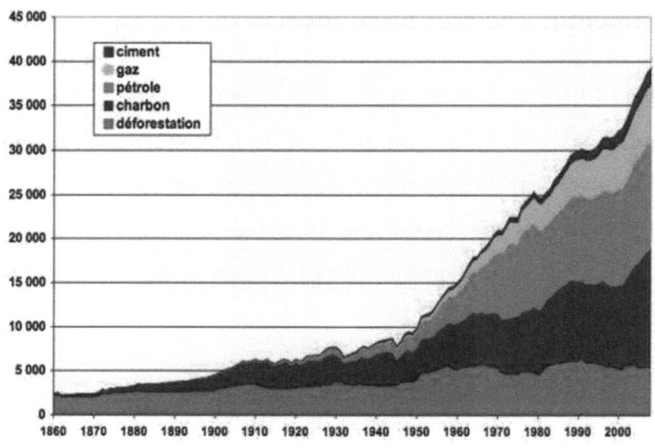

Source: J-M. Jancovici & sauvonsleclimat.org

Figure 1.4. Greenhouse gas emissions

5. Economic issues of fossil fuels in the transport sector

Since the industrial revolution, energy demand has been growing steadily (Figure 1.5). Global energy demand grew by 113% in 45 years, from 1973 to 2018; it was high in 2018. Its composition by activity is as follows: industry: 29%, transport sector: 29%, residential: 21%, tertiary: 8%, agriculture and fisheries: 2%, non-energy uses: 9%.

Among the most demanded energy sources, fossil fuels from oil are the most consumed (Figure 1.6). This behaviour demonstrates the critical dependence on these fuels, especially in the transport sector. In addition, the depletion of oil reserves presents a considerable problem for industrial advancement.

As a result, the search for cost-stable and sustainable fuels is a major imperative in the transport sector.

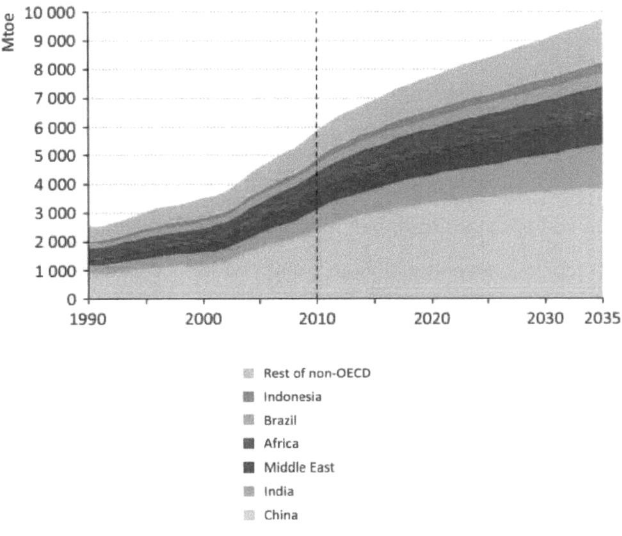

Source: World Energy Outlook 2012

Figure 1.5. World primary energy demand by region

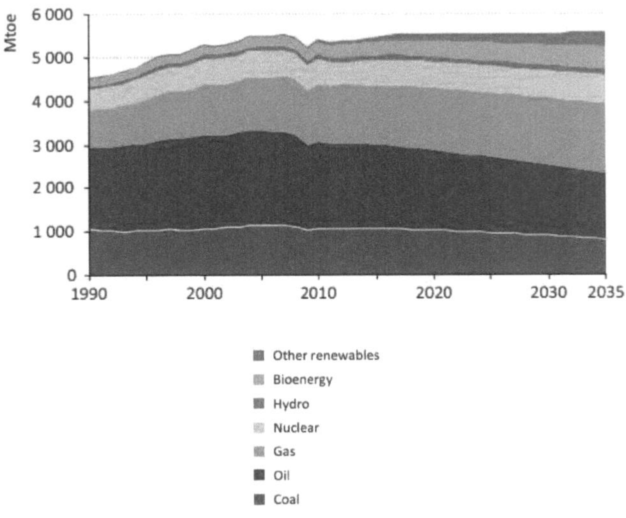

Source: World Energy Outlook 2012

Figure 1.6. Primary Energy Demand by Fuel

These economic and environmental challenges are forcing researchers and manufacturers of combustion engines for vehicles to review all the possibilities for reducing energy costs in new engines and those already in operation, and to replace conventional fuels with cleaner ones.

Gaseous fuels are the subject of renewed interest insofar as they are clean fuels, whose combustion reaction in the air results in minimal or even zero emissions at the engine's exhaust, compared to conventional liquid hydrocarbons. These alternative fuels are essentially in the form of liquefied petroleum gas (LPG), compressed natural gas (CNG) or hydrogen, and are therefore an environmental response adapted to the requirements of pollution control, particularly in urban areas.

Chapter 2: Alternative Fuels

1. Introduction

In the face of the enormous and aggressive exploitation of fossil energy resources, the global energy market is looking to explore alternative fuels specifically in the transportation sector for internal combustion engines (ICE). The pollution caused by the use of oil resources is increasing. The measures taken by the Kyoto Protocol show that CO_2 emissions have increased by 27%. Emissions from the combustion of fossil fuels and various industrial processes have contributed to the total increase in greenhouse gas emissions since the 1970s.

The overriding interest in alternative energy resources used in engines, observed at present, is the consequence of a patient concern for the environmental impact of ICMs, in particular the problems of greenhouse gases, and for the consumption of primary energy sources, which are limited. Emission regulations are becoming increasingly stringent to meet this challenge.

The terminology (alternative fuels) is generally used to identify energy sources that are not of fossil origin. They have specific characteristics related to a high calorific capacity and a combustion that results in lower emissions than conventional fuels. Another advantage of these fuels is the diversification of sources. The aim of this chapter is to characterise alternative fuels from the point of view of their resources, properties, production methods and application to combustion engines. The fuels that are used quite commonly can be classified into three categories: gaseous fuels (LPG, CNG...), biofuels (methanol, ethanol...) and hydrogen.

2. Gaseous alternative fuels

Typically, gaseous fuels generate very low levels of pollutants and can be used effectively in internal combustion engines. Gaseous fuels have wide ignition limits and can easily form homogeneous mixtures with air to allow complete combustion. Even very lean mixtures can be used. In addition, gaseous fuels have high hydrogen to carbon ratios, which has an impact on carbon emissions. Alternative gaseous fuels of potential interest for internal combustion engines are liquefied natural gas (LNG), compressed natural gas (CNG) and liquefied petroleum gas (LPG). Table 2.1 summarises the properties of LNG, CNG and LPG.

The production of these three gases is simplified in Figure 2.1.

Table 2.1: Physical and chemical properties of gaseous fuels

Properties	CNG	LNG	LPG
Density (under storage conditions) (kg/m3)	160	425	536
Calorific value* (MJ/kg or MJ/dm3)	4.80 (15°C, 20 MPa)	49.30 (- 150 ° C, 0.3 MPa)	46.05 (15 ° C, 1.5 MPa)
Auto-ignition temperature (° C)	540	540	450-510
Stoichiometric ratio (kg/kg)	17.1	17.1	15.7
Octane rating	105	105	110

Under the conditions of storage

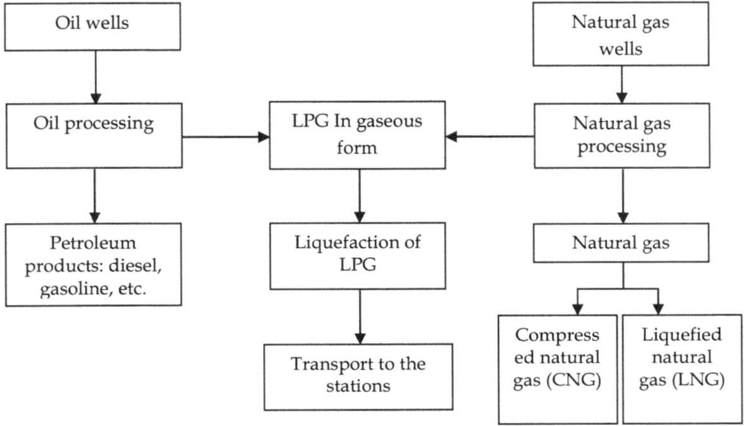

Figure 2.1. Production of gaseous alternative fuels

2.1 Liquefied petroleum gas

Liquefied petroleum gas used as a fuel is a hydrocarbon composed of C_3H_8 propane and C_4H_{10} butane, compressed (liquefied petroleum gases are derived from the lightest cuts in the distillation of petroleum.

2.1.1. Characteristics and specification of LPG

The general characteristics of LPG fuel are :

- The relative consumption of LPG fuel is reduced by almost *10% compared to* gasoline per unit volume.

- LPG has a higher octane rating of about 112, which allows the use of a higher compression ratio and gives a high thermal efficiency.

- Due to the gaseous nature of the fuel, the distribution between the cylinders is improved. Fuel consumption is also improved.

- The engine life is increased for the LPG engine since the use of cylinders is reduced.

This gaseous fuel has economic and environmental advantages (price incomparable with diesel and different physical and chemical properties ensuring lower emissions of exhaust pollutants). Research has shown that the operation of LPG gas engines emits significantly lower emissions than the operation of engines using traditional fuels.

In addition, LPG cars are rapidly developing as economical and ecological cars. In terms of safety, despite its volatile behaviour, LPG is known for its safe use. The fuel tank is generally much stronger than the tanks of conventional fuels.

2.1.2. LPG production

Generally, LPG is derived from two sources:

- ➢ gas fields (over 60%). The natural gas field supplies percentages of propane and butane. The natural gas is cooled to separate the different components. Butane and propane are also recovered during oil extraction as dissolved associated gases. The percentages of these gases in natural gas and crude oil vary greatly from field to field,

- ➢ petroleum refineries (Figure 2.2). During the refining of crude oil, butane and propane constitute between 2 and 3% of the total products obtained. They are the lightest cuts from the distillation of crude oil. These gases are also recovered from "secondary" processing operations, after the distillation phase.

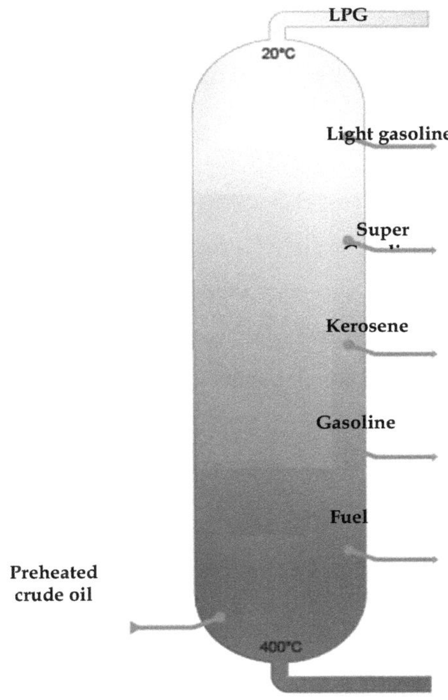

Figure 2.2. Production of LPG from petroleum refining

2.2. Natural gas

Natural gas is essentially composed of methane gas (CH_4). Its CO_2 emissions are much lower than those of conventional fuels. It exists in two known forms: CNG and LNG.

CNG has exactly the same properties as natural gas in its gaseous state. However, it can be stored at 300 times atmospheric pressure and transported in high-pressure tanks. It is supplied to the engine via a pressure regulator, either in the intake manifold or by direct injection into the cylinder.

Compressed natural gas engine technology significantly reduces vehicle and truck noise compared to trucks running on conventional fuels.

Liquefied natural gas LNG is composed of methane nevertheless also designs percentages of ethane, propane and butane. Methane is liquefied at a temperature of - 161°C at atmospheric pressure. In this form, LNG has a density of 422.62 kg m $^{-3}$.

2.2.1. Characteristics and specification of natural gas for vehicles

Generally, natural gas fuel for vehicles is characterized by:

> ➤ explosiveness: when dispersed in air, if its concentration is between 5 and 15%, natural gas burns or explodes on contact with a flame,
> ➤ The combustion of natural gas fuel is slower than that of other hydrocarbons and allows a significant reduction in vibrations and, consequently, in the noise level of the engines.
> ➤ CO_2 emissions: reduced by 25% compared to petrol
> ➤ particulate emissions: low to nil.

2.2.2. Natural Gas Development

The main steps in the development of natural gas are as follows:

> ➤ production: extraction at gas sites and platforms (Figure 2.3),
> ➤ transportation: gas transport through pipelines or specific vessels (Figure 2.4),
> ➤ storage: storage of gas in reservoirs or in old fields (Figure 2.5),
> ➤ distribution: sending gas to the operators

Sources: connaissancedesenergies.org

Figure 2.3. Natural Gas Production

Sources: connaissancedesenergies.org & futura-sciences.com

Figure 2.4. Natural Gas Transportation

Sources: connaissancedesenergies.org

Figure 2.5. Natural gas storage

3. Biofuels

Biofuel is a liquid or gaseous fuel made from non-fossil organic matter, derived from biomass, and is a substitute for conventional fuels.

3.1. Biogas

Biogas is a gaseous fuel produced by the fermentation of biomass (or methanation) of organic matter in an oxygen-free environment, see Figure 2.6.

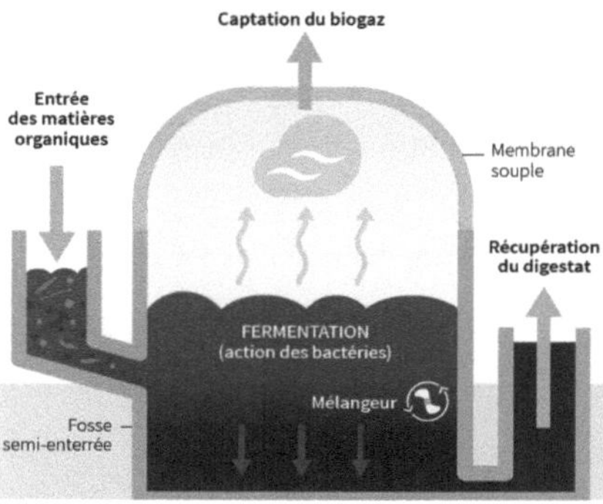

Sources: objectif2030.org

Figure 2.6. Fermentation procedure

3.1.1. Biogas characteristics

Biogas consists of more than 50% CH4 and contains almost 30% CO $_2$ and small amounts of hydrogen sulphide, water and other substances. In nature, each type of biogas has its own composition, which varies according to its organic source.

Biogas has many specifications, including

- ➢ Reduction of greenhouse gas emissions,
- ➢ Substitute for other fossil fuels,
- ➢ Control of the cost of the entire waste treatment,
- ➢ Economic and energetic valorisation of vegetable or animal waste,
- ➢ Participation in the hygienization of agricultural extracts,
- ➢ Reduction in the amount of chemical fertilizers used.

3.1.2. Biogas production

Biogas can be produced by different sources of biomass:

- ➢ Methanisation of raw plant organic substances (agricultural, household or food industry waste)
- ➢ Animal manure (manure, slurry)
- ➢ Lake and marsh bottoms
- ➢ Sludge and grease from wastewater treatment plants
- ➢ Landfills

In order to be used for the production of automotive biofuel, biogas must be "purified", i.e. enriched with methane, to ensure a quality that meets global pollution criteria. The technique is based on the separation of methane from CO_2 gas (biogas purification). Organic waste can be stored in a digester (biogas production reactor).

3.2. Biodiesel

Biodiesel is an alternative fuel for conventional diesel engines. Biodiesel can be used alone in self-ignition engines (B100 as 100% biomass) or mixed with diesel fuel (B30, B20,...). This biofuel is made from vegetable oil or animal oil transformed by a chemical process by mixing this oil with an alcohol (methanol or ethanol), figure 2.7.

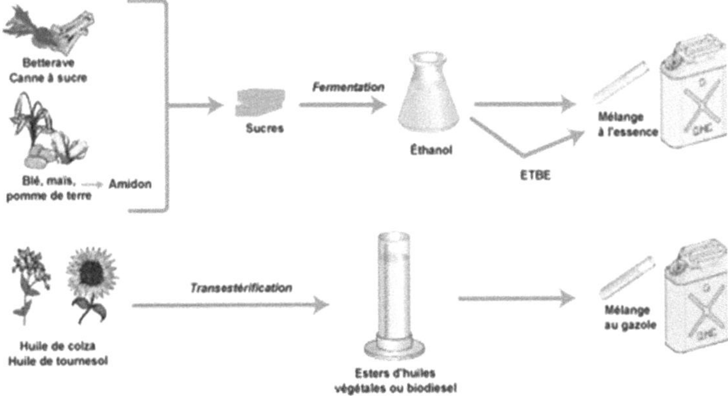

Source: IFP Énergies nouvelles

Figure 2.7. Biodiesel formation process

3.2.1. Characteristics of biodiesel

Compared to conventional diesel, vegetable oils have a higher viscosity. It is made from plants containing oil that has undergone chemical processing, such as soya, palm, etc. The oil must be mixed with methanol to obtain the VOME (vegetable oil methyl ester) mixture.

3.2.2. Production of biodiesel

The classic technique for the synthesis of biodiesel is called transesterification. It is a way in which vegetable oils or animal fats are mixed cold with an alcohol (ethanol or methanol) in the presence of a catalyst (sodium or potassium hydroxide), Figure 2.8.

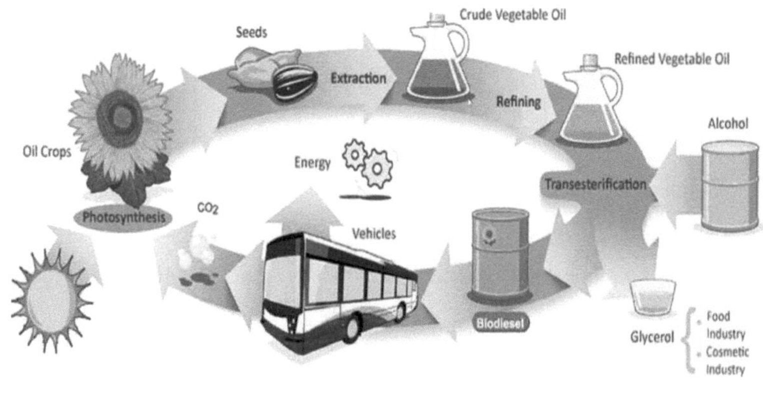

Source : eufuel.de

Figure 2.8. Biodiesel production

Chapter 3: Hydrogen Fuel

1. Introduction

Hydrogen is the most abundant element in the universe: 75% by mass and over 90% by number of atoms. On Earth, hydrogen is a component of water and living matter.

In the lexicon of combustion, the term hydrogen leads us directly to the component H2 (dihydrogen).

Hydrogen can be produced from a variety of feedstocks, including natural gas, biomass and water. Production techniques include steam reformation of natural gas, which is now the most economical process. Hydrogen can be used as a fuel in two ways: either with fuel cell technology or in an internal combustion engine where emissions are greatly reduced.

Hydrogen's contribution to the greenhouse effect is insignificant compared to other fuels since its combustion does not result in toxic components such as HC, CO and CO_2.

2. Properties of hydrogen

Hydrogen, made up of H2 molecules, is a colourless and odourless gas; it is the lightest of all bodies, its density compared to air being 0.07. It passes through porous walls. It is a gas difficult to liquefy.

Hydrogen produces two to three times more energy than other conventional fuels. It combines easily with oxygen, releasing considerable energy in the form of heat.

The wide flammability limit of hydrogen allows a wide range of engine power with the change in the equivalence ratio of the mixture. The flammability limit varies with pressure.

Table 3.1 shows a comparison of the characteristics of hydrogen with some other fuels.

Table 3.1. Characteristics of hydrogen via other fuels

Properties Hydrogen LPG Gasoline	Hydrogen	LPG	Gasoline
Formula	H_2	C_3H_8	C_8H_{18}-C_7H_{16}.
Auto-ignition temperature (°K)	858	723	530 – 730
Minimum ignition energy (MJ)			
Stoichiometric air-fuel ratio on a mass	0.02	0.25	0.8
basis	34.2	15.5	14.7
Density (kg/m3) @ CSTP			
Lower heating value (MJ/kg) @ CSTP			
Flame speed (m/s)	0.083 [gas]	500	750
Octane rating	120	45.7	43.9
	2.65 – 3.25	0.32	0.37 – 0.43
	130	111.5	80 - 92

CSTP: Standard Conditions of Temperature and Pressure

3. Hydrogen production

The production of H_2 is often achieved by the chemical extraction of fossil hydrocarbons. The electrolysis of water is also an industrialized process for the production of hydrogen. We quote thus:

➢ Steam reforming of natural gas. This involves reacting methane with water to obtain a mixture containing hydrogen and CO_2.

➢ The electrolysis of water. The electrolyser separates a water molecule into hydrogen and oxygen. This method is still not very widespread because it is much more expensive (2 to 3 times more expensive than natural gas reforming) and is currently reserved for specific uses, such as electronics, which require a high level of purity, figure 3.1.

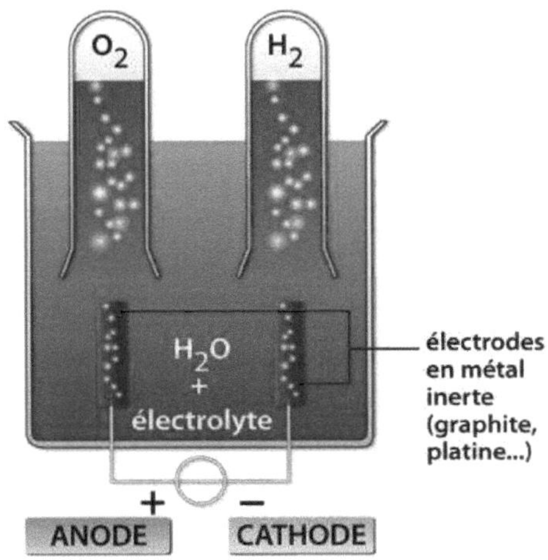

Figure 3.1. Electrolysis technique

Hydrogen produced today by steam-reforming methane costs about €1.5/kg H2. Dihydrogen produced from natural gas is the cheapest process.

Biomass is also an important source of dihydrogen: a mixture (CO + H2) is obtained by gasification and then purified.

Generally, there are other techniques for producing hydrogen. They are listed in Figure 3.2.

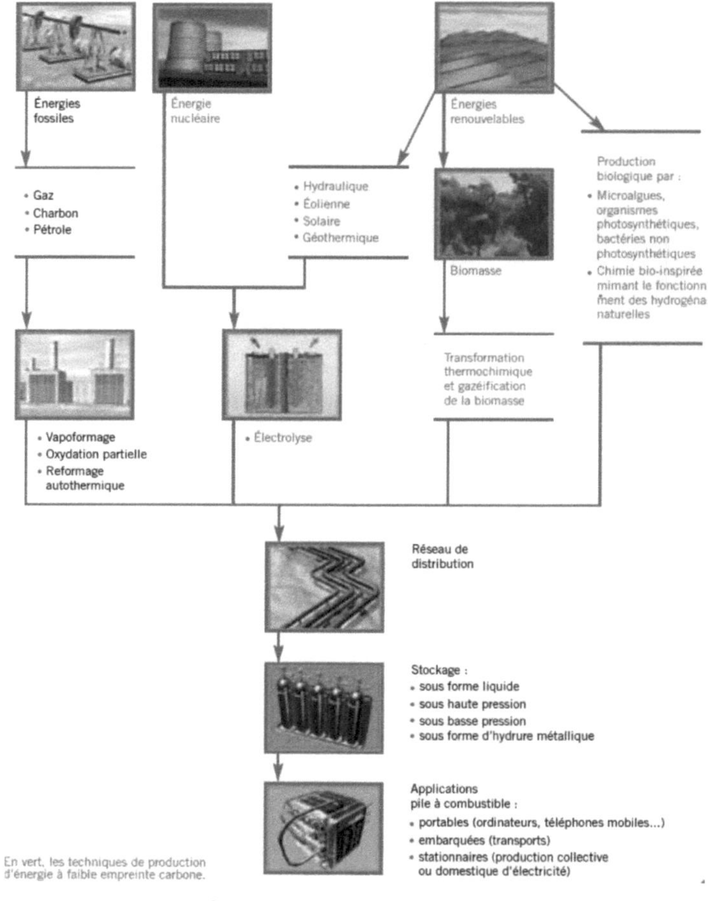

Source: Hydrogen, an energy carrier 2013

Figure 3.2. H2 production techniques

4. Hydrogen fuel technologies

4.1. Fuel cells

A fuel cell is based on the generation of an electrical voltage due to the oxidation on one electrode of a reducing fuel (e.g. H2) coupled with the reduction on the other electrode of an oxidant, such as oxygen from the air.

Fuel cells have multiple advantages:

- ➢ a high degree of modularity allowing a wide range of assemblies,
- ➢ Powerful flexibility of location regardless of the place of use,
- ➢ easy maintenance due to the absence of moving parts,
- ➢ an almost total absence of pollutant release such as CO_2,
- ➢ high efficiency of electricity generation,
- ➢ self-sufficient performance of the equipment load,
- ➢ short response times,
- ➢ very quiet operation.

The only significant obstacle is the cost.

The central part of a fuel cell is an electrolyte membrane (figure 3.3.). The first side is called the anode. On the second side, the cathode, is a positive electrode attracting electrons.

The operation of this battery is as follows:

- ➢ The hydrogen supplied to the cell enters the cell through the anode side. A platinum layer helps to split the hydrogen into electrons and protons.
- ➢ The electrolyte membrane has a proton pathway, but prevents the electrons from flowing, which are collected separately on the first side to deliver the electric current.
- ➢ As the oxygen travels towards the cathode, it meets another layer of platinum, which causes the oxygen and electrons to combine, producing water and heat.

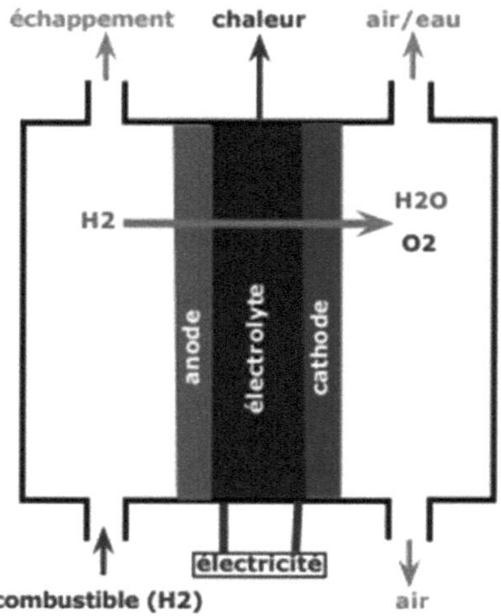

Figure 3.2. Fuel cell technology

In the case of vehicles, hydrogen is stored in dedicated tanks under pressure. The H2 and oxygen are fed to the fuel cell. These two gases undergo an electrochemical reaction in the cell. This reaction results in electricity, heat and water vapor.

4.2 Hydrogen in internal combustion engines

In internal combustion engines, hydrogen can be used as a fuel either in a mixture with other fuels or as pure hydrogen.

For blending operation, a distinction can be made between mixed operation and bi-fuel operation. Mixed operation is based on combinations of hydrogen with one or more other gaseous fuels or gasoline. The mixture is fed to the engine using a single fuel injection system. With this technology, hydrogen is continuously used to improve the performance of gas engines. On the other hand, bi-fuel operation is based on the supply of hydrogen and liquid fuels (such as diesel) in separate supply systems.

Chapter 4: Internal combustion engine technology

1. Introduction

An internal combustion engine (ICE) is a heat engine that delivers mechanical energy (i.e. work) from heat, most often produced by the combustion of a fuel (gasoline, kerosene, coal or hydrogen). This mechanical energy can be used to propel a car, to fly an airplane, or to drive an alternator that produces electric current. Its efficiency, i.e. the work it provides in relation to the amount of heat it consumes, is around 30%. The conversion of heat into work involves several steps. First, the heat must be converted into a force. Secondly, this force must be converted into a reciprocating motion, which in turn is converted into a rotating motion through the crank mechanism.

The fuels used in ICMs are essentially composed of carbon and hydrogen. They are therefore hydrocarbons. They are also fuels, since they can produce heat (calories) through combustion.

Most often liquid, fuels are used as a source of energy in thermal engines (gasoline and diesel engines, aircraft engines, gas turbines). The most widespread fuels today are derived from petroleum. Oil has not always been used as a fuel. In fact, in the early days of the automobile industry, engineers tried other sources of energy, such as oils, pulverized coal powder, gas, etc. But oil, being practical and inexpensive at the time, gradually became the preferred fuel.

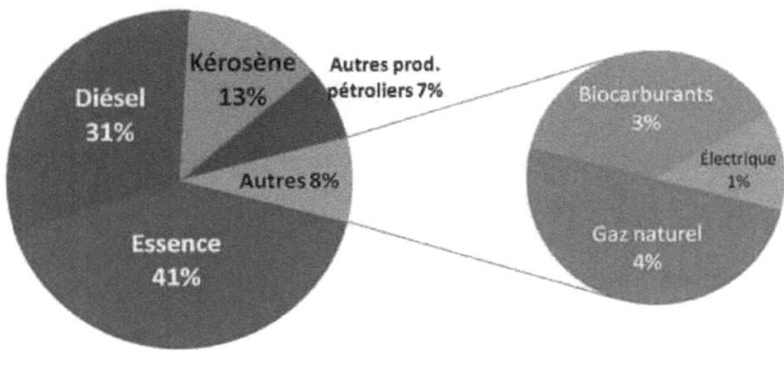

Source : afis

Figure 4.1. Fuels used in transportation

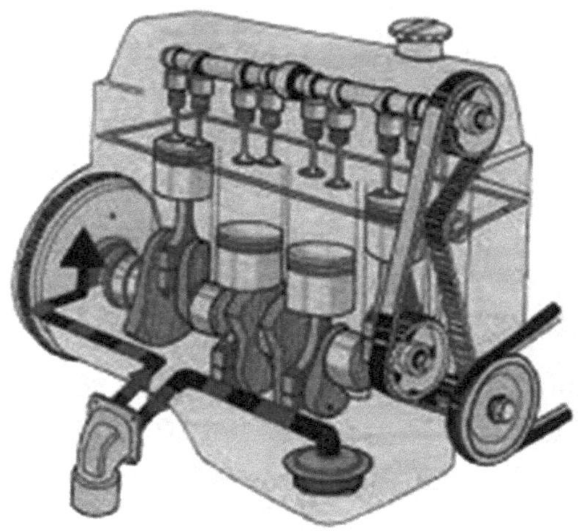

Figure 4.2. General schematic of the internal combustion engine

2. History

The history of the internal combustion engine goes through the following major chronological stages:

- 1700: steam engines.
- 1860: Lenoir's engine (efficiency $\eta\sim5\%$).
- 1862 Beau de Rochas defines the principle of the operating cycle of internal combustion engines.
- 1867: Otto's motor: ($\eta\sim11\%$ and rotation<90 rpm).
- 1876: Otto invents the 4-stroke spark ignition engine ($\eta\sim14\%$ and rotation< 160 rpm).
- 1880: Two stroke engine.
- 1892: Diesel invents the four-stroke compression-ignition engine.
- 1957: Wankel invents the rotary piston engine.

3. Classification of heat engines

Internal combustion engines can be classified into distinct families according to the nature of the fuel used, the operating technology and the ignition technique:

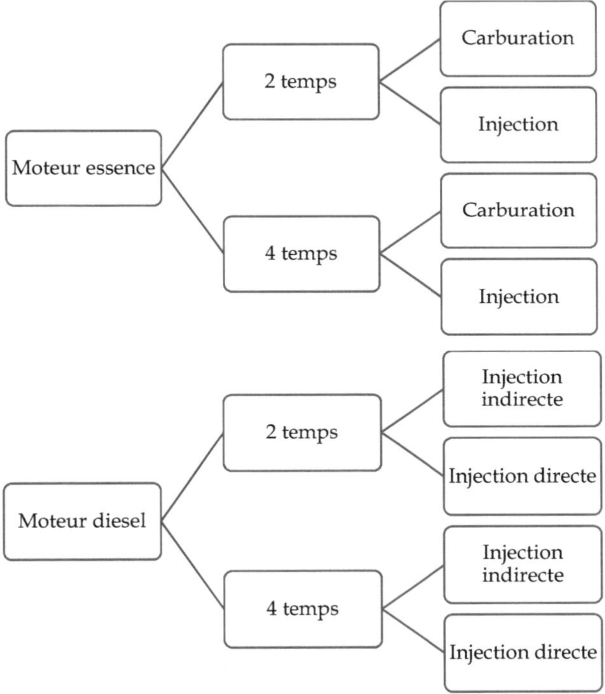

Figure 4.3. General classification of internal combustion engines

3.1 Classification by operating cycle

3.1.1. Four-stroke engines

The operating cycle can be broken down analytically into four times or phases. The movement of the piston is initiated by the combustion (rapid increase in the volume of gases) of a mixture of fuel and air (oxidizer) which takes place during the engine stroke. This is the only stroke that produces energy; the other three strokes consume energy but make it possible. The piston moves during the start by means of an external power source (often a starter or launcher: an electric motor is temporarily coupled to the crankshaft) until at least one stroke produces a force capable of ensuring the other three strokes before the next stroke. The engine then runs alone and produces torque on its output shaft.

The description of the successive cycles of a four-stroke engine is as follows:

- intake of a mixture of air and vaporized fuel, present in the intake pipe, mixture prepared by various components (carburetor or indirect injection system): opening of the intake valve and descent of the piston, the latter sucks this mixture into the cylinder at a pressure of -0.1 to -0.3 bar ;

- compression of the mixture: closing of the intake valve, then raising of the piston which compresses the mixture to 30 bars and 400 to 500°C in the combustion chamber;

- combustion (expansion around top dead center): the moment when the piston reaches its highest point and the compression is at its maximum; the spark plug, connected to a high-voltage electricity generator, produces a spark; the rapid combustion that follows constitutes the engine's stroke; the hot gases at a pressure of 40 to 60 bars push the piston back, initiating the movement;

- Exhaust: opening of the exhaust valve and raising of the piston which pushes the expanded burnt gases into the exhaust manifold, leaving room for a new charge of air/fuel mixture.

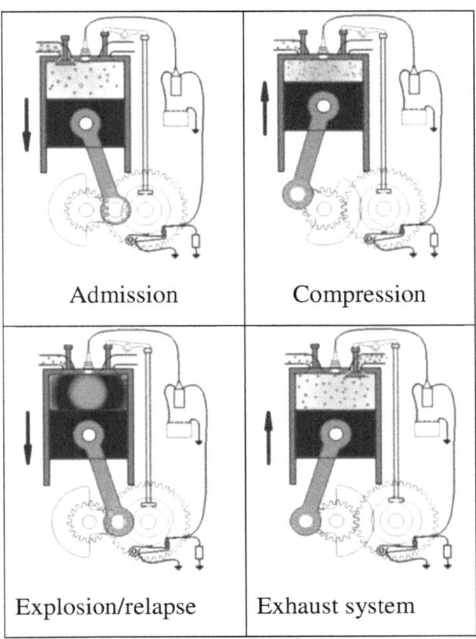

Figure 4.4. Cycles of the 4-stroke engine

3.1.2. Two stroke engines

Two-stroke engines are based on the movement and scavenging of both sides of the piston: the upper part for the compression and combustion phases and the lower part for the transfer of intake (and consequently exhaust) gases. This saves the movements (therefore latency, friction, etc.) of two non-energy producing cycles and produces more torque and power. The theoretical power of a 2-stroke engine is double that of a 4-stroke engine, but the fact of eliminating two strokes creates difficulties because the burnt gases must be expelled before admitting the air, and this in a very short time. Exhaust and intake must be carried out simultaneously in the vicinity of BDC (bottom dead centre) with the compulsory assistance of an air pressure higher than the atmospheric pressure provided either by a scavenging pump attached (alternative or rotary), or by a turbo blower.

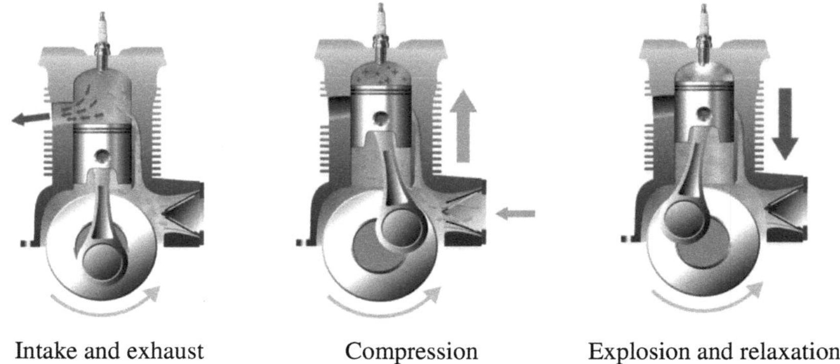

Figure 4.5. Cycles of the 2-stroke engine

3.2 Classification by ignition mode

3.2.1. Spark ignition engines (internal combustion engines)

A spark-ignition engine, more commonly known as a gasoline engine because of the type of fuel most commonly used, is a family of internal combustion engines, which may be reciprocating (two-stroke or four-stroke) or, more rarely, rotary (such as the Wankel engine).

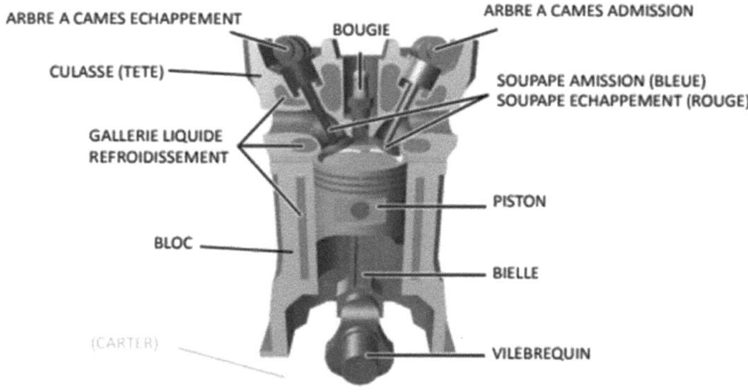

Figure 4.6. Spark ignition engine

In contrast to the diesel engine, the fuel mixture in a spark-ignition engine is not supposed to ignite spontaneously during operation, but by the action of a spark caused by the spark plug. It is therefore equipped with a complete ignition system, consisting of a spark plug, which causes the electric arc that ignites the gases in the combustion chamber, a coil to produce the high voltages necessary to create the spark, and an ignition control system (switch or electronic system).

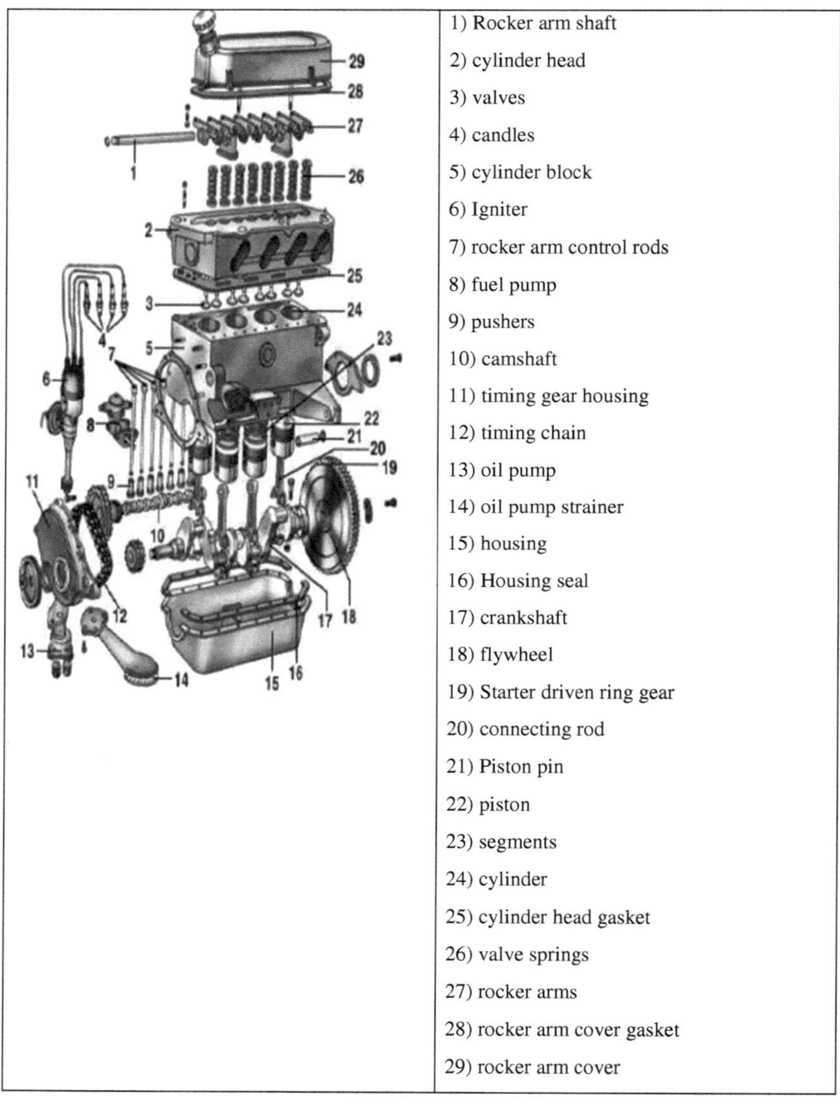

1) Rocker arm shaft
2) cylinder head
3) valves
4) candles
5) cylinder block
6) Igniter
7) rocker arm control rods
8) fuel pump
9) pushers
10) camshaft
11) timing gear housing
12) timing chain
13) oil pump
14) oil pump strainer
15) housing
16) Housing seal
17) crankshaft
18) flywheel
19) Starter driven ring gear
20) connecting rod
21) Piston pin
22) piston
23) segments
24) cylinder
25) cylinder head gasket
26) valve springs
27) rocker arms
28) rocker arm cover gasket
29) rocker arm cover

Figure 4.7. Exploded view of a spark ignition engine

3.2.2. Self-igniting engines (diesel engines)

The diesel engine, like the internal combustion engine, is an internal combustion engine. The architecture and operation of these two engines are similar. Their difference lies mainly in the way the fuel is burned.

In the internal combustion engine, gasoline is usually injected into the air and then this mixture is admitted into the cylinder (this is called indirect injection). This mixture, carefully metered, is then stirred and compressed. Combustion, triggered by the electric spark of a spark plug, then spreads by itself through the mixture until it is burnt completely. In the diesel engine, only a large amount of air is introduced into the cylinder. Once it is enclosed, it is strongly compressed to inject the diesel fuel into the combustion chamber.

3.3 Classification according to the arrangement of the pistons

There are several engine technologies related to the arrangement of pistons and cylinders. Among them, the most encountered are:

A- In-line engine: the cylinders are arranged vertically in the same plane.

B- V engine: the cylinders are divided into 2 equal groups according to 2 planes converging towards the crankshaft, angles of 90° and 120°.

C- Opposed flat engine: as in the V-engine, the cylinders drive the same crankshaft but opposite each other at 180°.

In-line motor Opposite motor

V-motor

Figure 4.8. Different piston arrangements

4. Different circuits in engines

4.1. Cooling system

Almost all engines have a cooling system to remove waste heat from the block and internal parts. The cooling system consists of a closed loop similar to that of a car engine and contains the following main components: water pump, radiator or heat exchanger, water jacket (which consists of ducts in the block and cylinder head) and a thermostat.

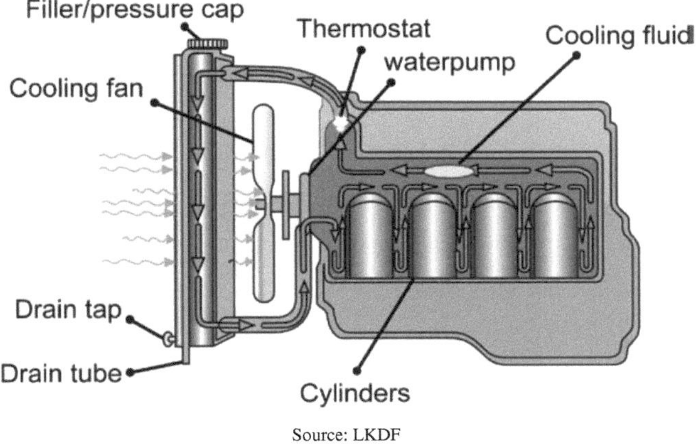

Source: LKDF

Figure 4.9. Cooling system

4.2 Lubrication system

An internal combustion engine would run for even a few minutes if the moving parts were allowed to make metallic contact. The heat generated by the enormous amount of friction would melt the metal and destroy the engine. To prevent this, all moving parts run on a thin film of oil that is applied between all moving parts of the engine.

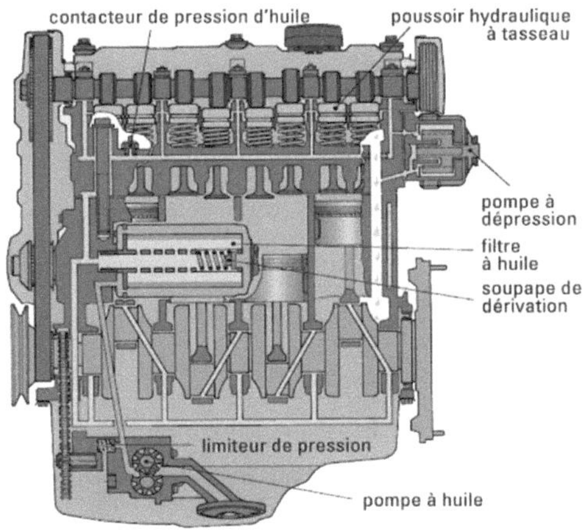

Figure 4.10. Lubrication circuit

41

4.3. Fuel supply system

4.3.1. Diesel engine

A system for storing and delivering fuel to the engine is required in all diesel engines. Diesel engines rely on injectors, which are precision components with very tight tolerance levels and very small injection ports, so the fuel delivered to the engine must be extremely clean and free of pollutants.

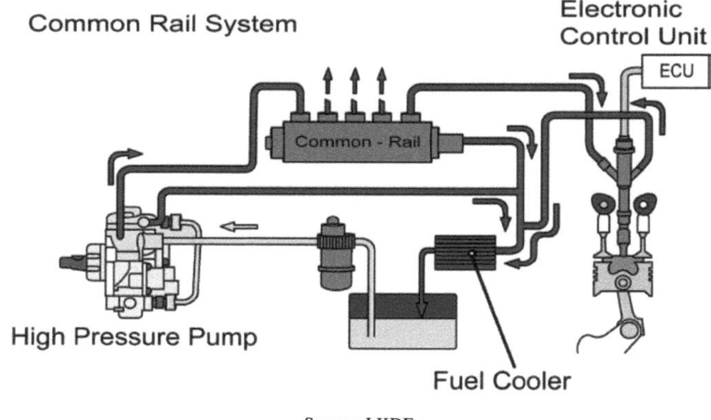

Source: LKDF

Figure 4.11. Diesel oil supply system for diesel engine

4.3.2. Gasoline engine

The purpose of the fuel system is to supply the engine with the air and fuel needed for proper combustion. The supply system consists of two different circuits: the air supply circuit and the fuel supply circuit. Two solutions are used to achieve the mixture:

➤ Carburetor system: the air-fuel mixture is obtained in the carburetor and then introduced into the engine cylinder.

➤ Injection system: the mixture is made in the intake pipe, the air is supplied by a conventional way and the petrol is injected under pressure by injectors (one per cylinder).

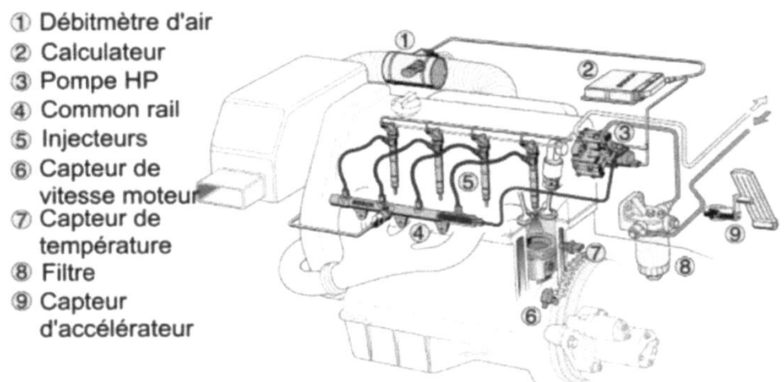

Figure 4.12. Direct fuel injection circuit and system

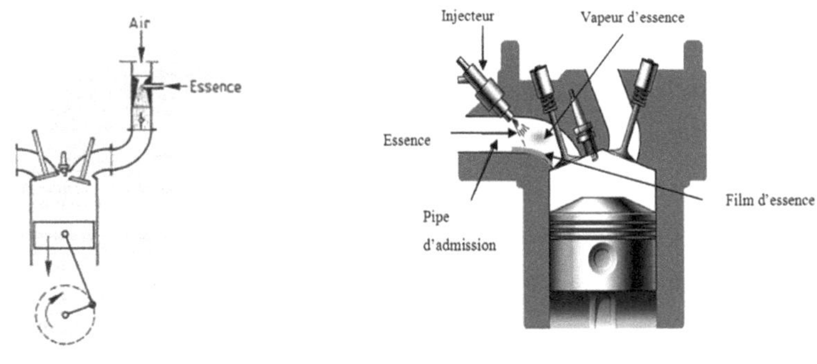

Gasoline fueling system

Indirect fuel injection system

Figure 4.13. Fuel supply system with indirect carburetion and injection

Injection systems can be classified according to where the fuel is injected into the air drawn in by the engine:

> ➤ The injection is direct if it is carried out in the combustion chamber of the cylinder.
>
> ➤ Injection is indirect if it takes place in the intake manifold, more or less close to the intake valve, with the fuel jet directed towards the valve.

➢ Centralized injection if it is done in the part of the manifold common to all the cylinders, in the place that a carburetor would occupy.

Injection systems can also be differentiated by the control device:

➢ In mechanical injection, the pump, which is mechanically driven by the engine, pressurizes the fuel and meters the injected volume.

➢ In electronic injection, the electric pump supplies the fuel under pressure; the functions of metering, regulation and injection are totally or partially controlled by an electronic control unit.

4.4 Ignition circuit

In a spark ignition internal combustion engine, ignition is the mechanism that initiates the combustion of the flammable gas in the combustion chamber of each cylinder. There are 3 basic circuits in the ignition system:

➢ The low voltage circuit which uses the current coming from the battery (6 V or 12 V) before converting it into high voltage in the coil.

➢ The high voltage circuit (15000 V) which takes the current from the coil (or electronic ignition system) and sends it to the spark plugs via the distributor.

➢ The circuit between the coil and the distributor capacitor, and the return wire (ground) to the battery terminal.

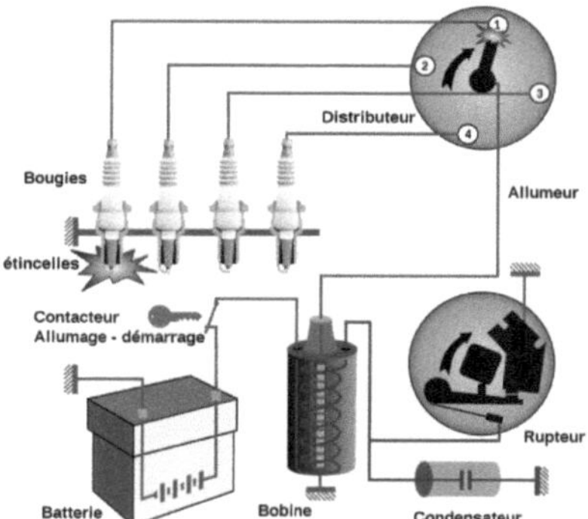

Figure 4.14. Ignition system (spark ignition engine)

5. Techniques for improving the performance of internal combustion engines (Turbocharging)

The turbo is part of the air intake system and is used to compress incoming fresh air and distribute it to the cylinders for improved combustion. The turbo can be part of a turbocharged or supercharged air intake system. More information on these types of turbos is given at the end of the module.

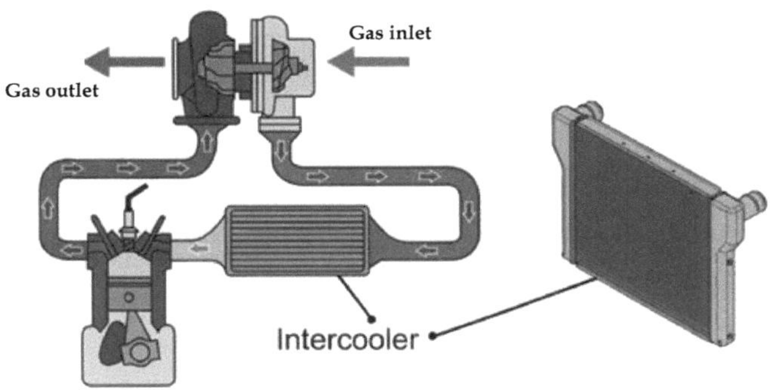

Source: LKDF

Figure 5.15. Turbocharger system

Turbocharging in an engine occurs when engine exhaust gases drive a turbine (pump) that rotates and is attached to a second pump located in the fresh air intake system. The intake system pump compresses the fresh air.

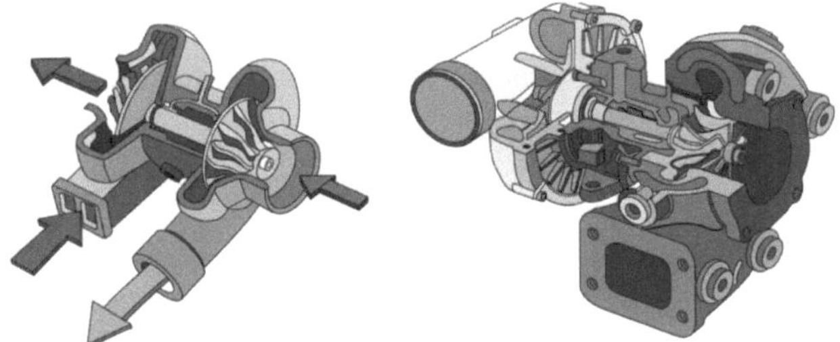

Figure 4.16. Components of the turbocharger

Compressed air has two functions. First, it increases the engine's available power by increasing the maximum amount of air (oxygen) that is pushed into each cylinder. Thus, more fuel can be injected and more power is produced by the engine. The second function is to increase the intake pressure. This improves the scavenging of air from the exhaust gases out of the cylinder. Turbocharging is usually available in powerful four-stroke engines.

It can also be used in two-stroke engines, where the increase in intake pressure generated by the turbocharger is required to force fresh air into the cylinder and help force exhaust gases out of the cylinder to keep the engine running.

6. Basic operational terminology for internal combustion engines

To understand the detailed operation of an internal combustion engine, several terms must be defined.

- BORE AND STROKE: The bore refers to the cylinder diameter of the engine, and the stroke refers to the distance the piston travels from the top of the cylinder downwards.

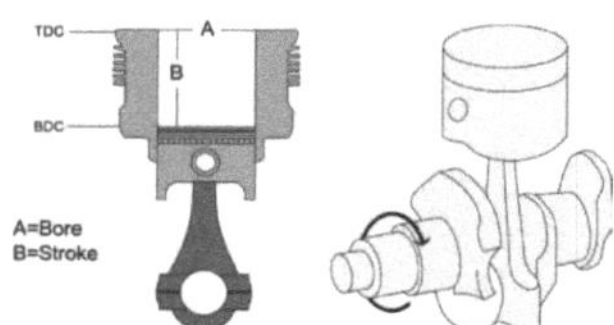

- CYLINDRIA: The displacement refers to the total volume displaced by all the pistons during a stroke

- COMPRESSION RATE AND DEAD SPACE VOLUME: The dead space volume is the volume remaining in the cylinder when the piston is at TDC. The compression ratio of the engine is determined by taking the volume of the cylinder with the piston at TDC and dividing the volume of the cylinder with the piston at TDC by the volume of the cylinder with the piston at TDC.

- IGNITION ORDER : The firing order is the order in which each cylinder in a multiple cylinders ignite (power stroke). For example, the firing order of a four-cylinder engine could be 1-3-4-2.

- ALTERNATOR: A device for producing electric current. An alternator produces alternating current rectified to direct current to charge the battery.

- BATTERY: A device for storing electrical energy. Its role is to activate the starter and accessories when the engine is initially started. The battery recharges automatically while the engine is running.

- CARTER, CASING: Name given to any component in the form of a housing or casing containing moving mechanical parts. Examples: engine casing, gearbox, differential casing.

- CAM SHAFT: The shaft that turns in the cylinder head. It has two cams per cylinder: one that controls the intake valve and one that controls the exhaust valve.

- CARBURATOR: A device for converting fuel and air into combustible gas. The proportion of fuel and air in this mixture should be as even as possible.

- CONNECTION ROD: The part that connects the piston to the crankshaft. The connecting rod transforms the vertical linear motion of the piston into rotary motion of the crankshaft.

- CRANKSHAFT: Part that receives the mechanical energy delivered by the pistons (via the connecting rods). The crankshaft is made of very hard (forged) steel.

- CYLINDER: The chamber in which the piston moves. It is hollowed out of the engine block.

- CUSHION: The upper part of the engine (attached to the top of the cylinder block) in which explosions occur. It contains the valves, spark plugs and camshaft. This part of the engine reaches a very high temperature and must be cooled continuously.

- DISPLACEMENT: The volume displaced by all the pistons in an internal combustion engine. This value is expressed in cm3 or liters.

- FUEL INJECTION: A carburetion system in which the carburetor is replaced by a fuel injection pump, as in a diesel engine. When the fuel injection system is electronically controlled, it delivers a much more precise amount of fuel to the engine when the vehicle is moving at high speeds.

- FUEL SYSTEM: The group of components that supply fuel and air to the carburetor. It includes: fuel tank, fuel filters, fuel pump, fuel supply lines and air filter.

- IGNITION SYSTEM: The group of electrical circuits required to ignite the air/fuel mixture drawn into the engine cylinders.

- EXHAUST ADMISSION SYSTEM: The group of mechanical parts that supply the air/fuel mixture and exhaust the burnt gases. It consists of the valves, rocker arms and camshaft with its gears.

- PISTON: A cylindrical metal part that moves up and down in the cylinder. The piston is connected to the crankshaft by the connecting rod. The movement of the pistons governs the four strokes of the cycle invented by Alphonse Beau De Rochas.

- CUTTER: A lever that swings on a central pivot to transmit the movement of the cams to the valves. Another part called a push rod acts as a link between the rocker arm and the camshaft. There are two rocker arms per cylinder.

- FIREPLUG: A device that triggers the explosion in the combustion chamber. It is part of the electrical system. It has two electrodes, one negative, connected to the ground through the engine block and the other positive connected to the distributor. These two points are separated by a small distance. When a high voltage (at least 14,000 volts) is applied to the + terminal, an electric arc is formed across the gap and ignites the air/fuel mixture in the combustion chamber. The spark plug sparks 10, 25 or even 50 times per second, depending on the engine speed.

- STARTER: A small electric motor powered by the battery when the ignition key is pressed. The starter turns the engine until it "takes off" (runs on its own power).

- VALVE: A part made of special steel and used to open and close the intake and exhaust ports. There may be two or four valves per cylinder. The valves are controlled by the camshaft and its mechanism.

Chapter 5: Alternative fuels in internal combustion engines

1. Introduction

Gas engines are usually manufactured to run directly on gas with a single fuel or they are derived from liquid fuel engines and are technologically adapted to burn the gas fuel. There are two types of adaptation of conventional engines: conversion of spark ignition engines with minor modifications and conversion of compression ignition engines with major modifications. In this chapter, a review of the technologies for adapting these engines is presented, as well as the criteria needed to convert an engine to gas under good operating conditions.

2. Technologies for engines running on gaseous alternative fuels

The use of gas (LPG, CNG and H2) as an alternative fuel in an engine can be divided into three main types according to their fuel consumption.

2.1. Single fuel dedicated engine

This is a specialized type of engine, which has been designed and optimized to run only on gas. This allows the characteristics of gas to be fully exploited without the need to compromise the engine design by consuming other fuels. This type of engine is primarily intended for heavy-duty vehicles such as marine applications or engines for power generators.

2.2 Dual-fuel engine

a. Dual fuel diesel-gas engine

This is a development of the conventional diesel engine. In this type of engine, diesel and gas are introduced into the engine cylinders at the same time during the compression phase. The first characteristic of a dual-fuel engine is the way in which combustion is initiated. The objective of the compression phase is to raise the temperature of the oxidant or oxidant-fuel mixture above its flammability limit. The second characteristic of the dual-fuel engine (also known as diesel-gas) is related to the nature of the fuels used, at least one of which must be gaseous. The technological characteristics of the engine, which derive from the two previous characteristics, concern the compression ratio, which must be sufficient to allow self-ignition, and the nature of the fuel, which may be gaseous or liquid.

As gas does not ignite on compression alone, the diesel fuel is required to act to ignite the air-gas mixture. When the gas supply is not available, the engine can revert to conventional diesel operation. Diesel is mainly used to heat the engine; at full throttle, the gas content averages 65 % to 85 %. The proportion of gas is constantly adapted to the power requirements of the engine. This combination of gas and diesel reduces the total amount of exhaust pollutants. The dual-fuel engine consumes considerably less energy than conventional engines. The gas injection is controlled by the diesel injection.

b. Hydrogen-gas bi-fuel engine (LPG or CNG)

The main motivation for adding hydrogen to a gas is to extend the lean burn limit of the gaseous fuel. On the other hand, the low storage density of compressed hydrogen tanks can be significantly improved by mixing hydrogen with gases.

3. Technologies for converting conventional engines to alternative fuels

3.1 Diesel engine conversion

Generally, the steps for converting a diesel engine to LPG, CNG or hydrogen gas are carried out as follows:

➢ Reduction of compression ratio from the original value to a value corresponding to spark ignition operation (usually less than 12) by reducing piston sizes.

➢ Add a new air-fuel mixture supply system.

➢ Installation of gas components. Gases must be stored in a cylindrical tank at an adequate pressure. The regulator, solenoid valve, safety elements and injectors must be properly installed.

➢ Installation of the ignition system.

➢ Replacement of the air intake manifold with one suitable with the air-fuel mixture flow. The distribution system characteristics of the existing diesel version are adopted on the spark ignition engine without modification.

In this type of converted engine, the mechanical performance is generally reduced (Figure 5.1), but the benefit is in the reduction of toxic emissions and fuel consumption (Figure 5.2).

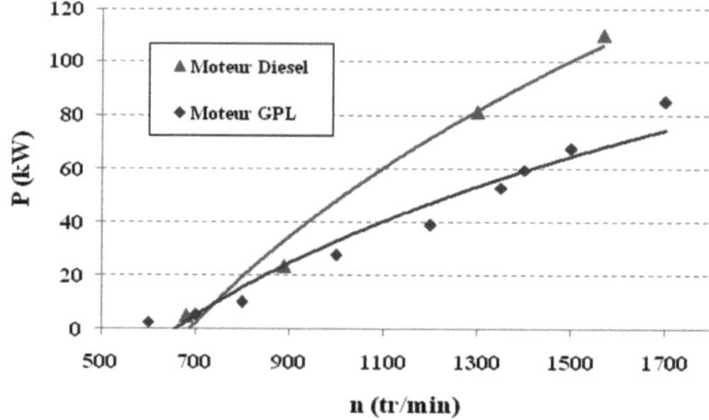

Figure 5.1. Power of a diesel engine converted to LPG

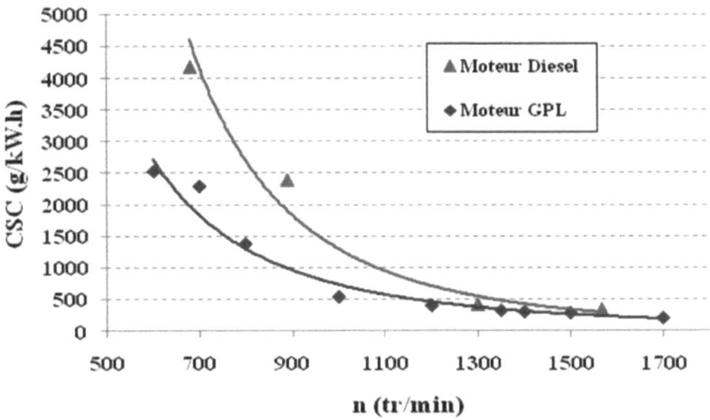

Figure 5.2. Specific consumption of a diesel engine converted to LPG

3.2 Conversion of gasoline engines

Converting a gasoline engine to petroleum gas, natural gas or hydrogen does not change the operating principle of the engine, which is spark ignition, only a few modifications are required:

> ➢ Installation of injectors adapted to the operation of gaseous fuel. The injection can be single point indirect injection, multipoint indirect injection or direct injection

> ➢ Installation of an intake system tuned to the supercharging technology is possible since the hydrogen occupies a relatively large volume, which reduces the quantity of working fluid in each cycle and significantly reduces the specific power of the engine. The limitation in specific power can be eliminated by appropriate injection devices (high pressure direct injection).

> ➢ Install a fuel supply system that is safe from self-ignition and flashback (Figure 5.3). Hydrogen, for example, is very susceptible to both of these problems.

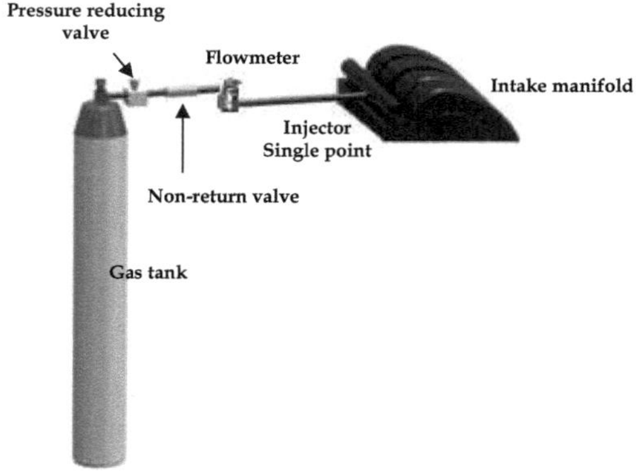

Figure 5.3. Gaseous fuel supply system

Similarly, in this type of converted engine, mechanical performance is generally reduced slightly (Figure 5.4), but the benefit is in the reduction of toxic emissions and fuel consumption (Figure 5.5 and Figure 5.6).

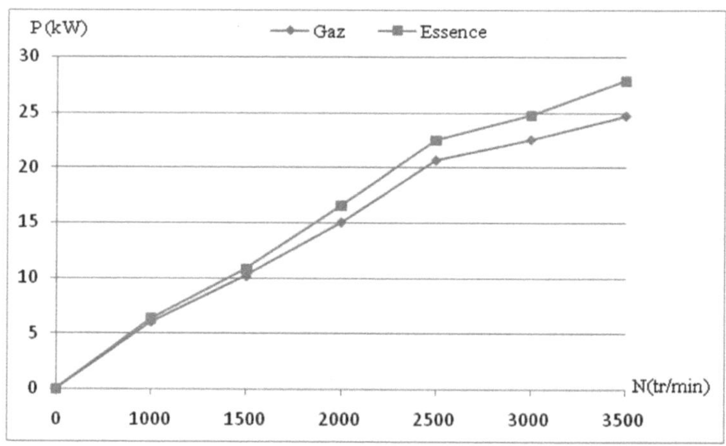

Figure 5.4. Power of a petrol engine converted to LPG

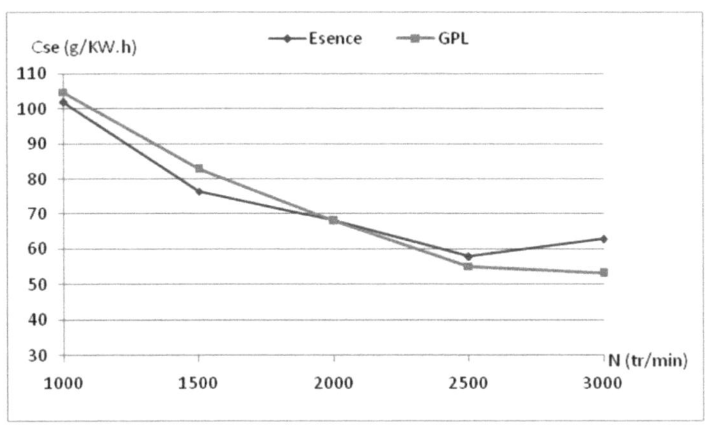

Figure 5.5. Specific consumption of a petrol engine converted to LPG

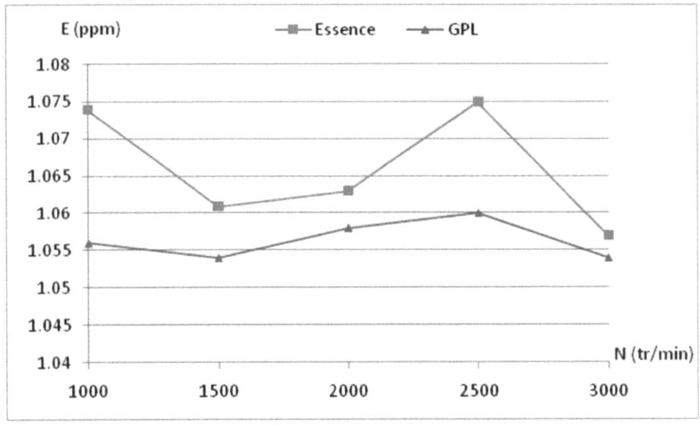

Figure 5.6. Comparison of NOx emissions from a petrol engine converted to LPG

Chapter 6. Future and challenges of hydrogen and gaseous fuels use in transport

1. Introduction

As part of its objectives to reduce fossil fuel consumption and toxic emissions, the land transport sector is committed to moving to a cleaner energy scenario, based on regulating energy consumption and reducing pollutant emissions. Hydrogen and other gaseous fuels (CNG, LPG and biogas) have some appeal for their properties as alternative fuels.

Electric and hybrid vehicles offer solutions that can also be used to solve economic and ecological problems. However, their drawbacks, which have been increasing in recent years with regard to range and safety, are forcing car manufacturers to improve the performance of internal combustion engines using hydrogen and alternative fuels.

2. Hybrid vehicles

The hybrid car (Figure 6.1) is the best known of the alternative-powered cars. There are usually two engines in a hybrid vehicle; a combustion engine and an electric traction motor.

A hybrid propulsion is based on the combination of a combustion engine with an electric one, in order to benefit from the advantages of each type of engine: the flexibility of the traction engine and the high power and autonomy of the combustion engine.

2.1. Hybrid vehicle components

The hybrid car is essentially made up of :

> ➤ A combustion engine connected to the vehicle's transmission. It operates on gasoline, diesel or bio-fuel. It allows the propulsion of the vehicle and the recharge of the battery.

➢ An electric motor that shares the role of propulsion with the combustion engine.

➢ A battery that stores the electricity created by the combustion engine and the energy from braking, and transfers it to the electric motor.

➢ A control and command unit that manages the various parts of the vehicle.

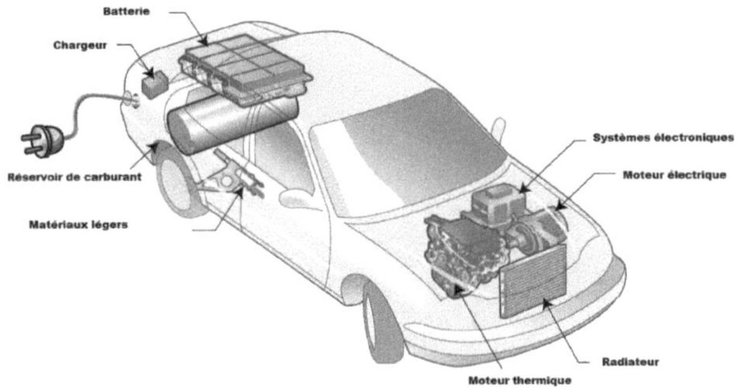

Source : futura-sciences

Figure 6.1. Hybrid vehicle

2.2 Disadvantages of the hybrid vehicle

Despite the advantages of hybrid vehicles in terms of reduced emissions, they have considerable disadvantages:

➢ The purchase cost of hybrid vehicles is higher than that of conventional vehicles.

➢ Batteries degrade and cannot be used indefinitely. The concern is that they are expensive.

➢ The disproportionate consumption of fossil fuels in the battery manufacturing process, and the emissions associated with this process.

➢ The safety of the batteries, since the risk of ignition of the vehicle increases compared to what works with combustion engines.

➢ The high voltage of the batteries in a hybrid car is potentially dangerous in the event of an accident.

3. Electric vehicles

The electric vehicle is an automobile powered by an electric motor driven by a storage battery (Figure 6.2). Tesla is the company most involved worldwide in the manufacture of this type of vehicle.

The market share of electric vehicles reaches almost 4.1% of global sales in Q1 2021.

Figure 6.2. Electric vehicle

3.1. Principle of electric vehicles

The electric vehicle (Figure 6.3) basically consists of a battery system and one or more electric motors. It differs from a vehicle powered by an internal combustion engine in that it takes longer to recharge.

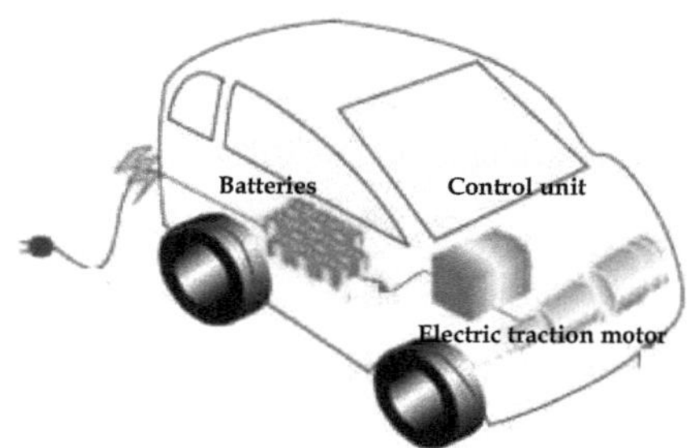

Figure 6.3. Components of the electric vehicle

3.2. Benefits of electric vehicles

The known advantages of electric vehicles are:

> The electric car is specified by zero emissions. By comparing it with thermal cars that generate toxic emissions of CO_2, nitrogen oxide and particles, electric motors are systematically considered as a method to curb the problem of global warming.

> Noise is reduced. The electric vehicle engine generates a lower noise level than the combustion engine because of the absence of the moving parts of combustion engines.

> The electric vehicle, which does not consume fuel, offers the possibility of reducing the amount of oil consumed

3.3 Disadvantages of electric vehicles

> The size of the vehicle and the distribution of the mass of the functional parts.

> The ageing and life span of batteries is not well known.

> The enormous consumption of conventional fuels in the battery manufacturing process, and the emissions associated with this process.

> The safety of the batteries, at least for some types of lithium models, is not yet fully guaranteed.

> The traditional recharging time (approximately up to 8 hours for a 600 km journey) implies a new obligation to control the consumption time, the type of power supply and the rapid availability of the vehicle.

> The short range is the biggest drawback of the all-electric vehicle. Fuel consumption is extremely complex to evaluate because it varies according to several factors (driving style, speed, road gradient, etc.).

4. Future of hydrogen and gaseous fuels in transport

Due to the major disadvantages of battery-powered cars as stated above, the use of alternative fuels and in particular hydrogen presents the most reliable solution in the transport sector to obtain economic and ecological benefits.

The enormous advances in fuel technology play an important role in the exploitation of these gases in internal combustion engines.

4.1 High pressure injection of gaseous fuels and hydrogen

In the technology of spark ignition engines working only with gaseous fuels (LPG, CNG and hydrogen) the technology currently adopted to increase the energy performance of these engines is the direct injection under high pressure (figure 6.4). This technique is controlled in a valuable way to capture the maximum energy and calorific efficiency of the fuel.

Source : h2-mobile.fr

Figure 6.4. High pressure injection of H2

4.2. Arrangement of the injectors in the bi-fuel combustion chamber

In diesel engines, the technology of bi-fueling with another gaseous fuel in addition to diesel is currently very popular for its ecological and energy benefits. The arrangement of the injectors of the two fuels in this type of engine is a significant research topic to study the importance of the jets of diesel and alternative gas in the combustion chamber (figure 6.5).

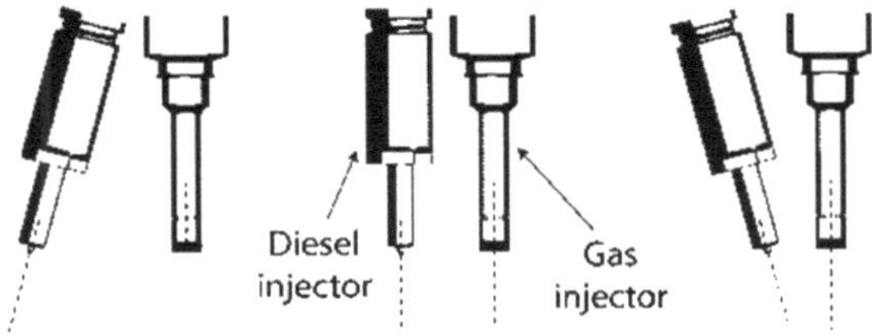

Figure 6.5. Diesel and gas injector arrangement

4.3. Safety techniques for the storage of gaseous fuels

Vehicles running on natural gas, LPG or hydrogen have a gas tank that must be well secured. Researchers in the transport sector have advanced their work with the aim of having a good functioning of the combustion engine with the totality of safety especially with the important storage pressures of H2 and CNG. The higher the storage pressure in the tanks (several hundred bars), the greater the quantity of gas delivered. For safety reasons, these composite tanks (Figure 6.6) are equipped with various additional features, including a solenoid valve, a high-pressure regulator and a pressure relief valve.

In case of hydrogen supply, H2 detectors are distributed in the vehicle and a flame arrestor is installed in the supply circuit.

Source : h2-mobile.fr

Figure 6.6. H2 gas tank and supply system

Bibliographic references

1- Benjamin Dessus. Transport and the challenges of energy and climate. Agence Française de Développement, 2009.

2- Olivier Apper, Philippe Pinchon. The cleaner, multi-energy car of the future. Press conference. IFP, 2004.

3- European Environment Agency, "Trends and projections in Europe 2017 Tracking progress towards Europe's climate and energy targets", 17/2017, 2017.

4- Core Writing Team, R.K. Pachauri and L.A. Meyer (eds.), "Climate Change 2014: Synthesis Report. Contribution of Working Groups I, II and III to the Fifth Assessment Report of the Intergovernmental Panel on Climate Change", IPCC, 2014,

5- Worldwide Look at Reserves and Production. Oil & Gas Journal, December 2009.

6- LAGRÉE Stéphane. The challenges of energy transition in Việt Nam and Southeast Asia. Knowledge Publishing House. 2018

7- Enerdata. World energy balance, 2021 edition

8- All about CNG. Product Marketing, TOTAL Marketing & Services. 2016

9- World energy outlook. International energy agency. 2012

10- Jean-François Marcoux. Savings with CNG and RNG in vehicle fleets. Gaz Métro. 2018

11- Axegaz. Natural gas production.

12- Abdeen Mustafa Omer. An overview of Biomass and Biogas for energy generation: Recent development and perspectives. Journal of Bioscience and Biotechnology Discovery. Article Number: 5CFE1FC02, 2016

13- Chintala V, Subramanian KA. Experimental investigation of hydrogen energy share improvement in a compression ignition engine using water injection and compression ratio reduction. Energy Conversion and Management 108: 106-119. 2016

14- Roger Cadiergues. Fuel cells. MemoCad nB42.a.

15- Jean Syrota, Philippe Hirtzman, Dominique Auverlot. The car of tomorrow: fuels and electricity. Centre d'analyse stratégique, 2010.

16- All about oil. IFP énergies nouvelles.

17- Ho Lung Yip et al. A Review of Hydrogen Direct Injection for Internal Combustion Engines: Towards Carbon-Free Combustion. Appl. Sci. 2019, 9, 4842.

18- Pierre Duysinx. Internal combustion engines. University of Liege, 2010.

19- Diesel engine. Introduction to mechanics. Vienna International Centre.